THE PORTAGE POETRY SERIES

SERIES TITLES

Dear Lo
Brady Bove

Sadness of the Apex Predator
Dion O'Reilly

Do Not Feed the Animal
Hikari Miya

The Watching Sky
Judy Brackett Crowe

Let It Be Told in a Single Breath
Russell Thorburn

The Blue Divide
Linda Nemec Foster

Lake, River, Mountain
Mark B. Hamilton

Talking Diamonds
Linda Nemec Foster

Poetic People Power
Tara Bracco (ed.)

The Green Vault Heist
David Salner

There is a Corner of Someplace Else
Camden Michael Jones

Everything Waits
Jonathan Graham

We Are Reckless
Christy Prahl

Always a Body
Molly Fuller

Bowed As If Laden With Snow
Megan Wildhood

Silent Letter
Gail Hanlon

New Wilderness
Jenifer DeBellis

Fulgurite
Catherine Kyle

The Body Is Burden and Delight
Sharon White

Deirdre Lockwood's *An Introduction to Error* is a collection governed by a powerful and threatening fragility. Full of poems that live at the edge of all those things—his heart, her body, the temperature of the planet and the pH of the ocean—we might finally be ready to admit we can't control, their lines are nevertheless measured in a scientist's careful music. For if love were an experiment we could execute in a lab, then these poems would be the notes: clear-eyed, detailed, and, above all else, responsible to their irresistible and intrinsic truth.

—KEETJE KUIPERS
author of *Lonely Women Make Good Lovers*

An Introduction to Error is a work of textural delicacy and tensile strength. These poems contain beautifully meticulous observations, from the subtle wonders of "parcels of water" to a "crimson sleeve whirling in the wash." Lockwood's poetic voice has the rigorous grace of scientific methodology coupled with the tenderness and patience to allow each poem to work its own elegant calculus of measure and balance.

—LAURA DA'
Washington State Book Award winner
author of *Severalty*

AN INTRODUCTION TO ERROR

POEMS

DEIRDRE LOCKWOOD

CORNERSTONE PRESS
UNIVERSITY OF WISCONSIN-STEVENS POINT

Cornerstone Press, Stevens Point, Wisconsin 54481
Copyright © 2025 Deirdre Lockwood
www.uwsp.edu/cornerstone

Printed in the United States of America.

Library of Congress Control Number: 2025941892
ISBN: 978-1-968148-15-7

Cornerstone Press titles are produced in courses and internships offered by the
Department of English at the University of Wisconsin–Stevens Point.

DIRECTOR & PUBLISHER   EXECUTIVE EDITORS
Dr. Ross K. Tangedal   Jeff Snowbarger, Freesia McKee

EDITORIAL DIRECTOR   SENIOR EDITORS
Brett Hill   Paige Biever, Eva Nielsen, Reilly Crous

PRESS STAFF
Brianna Loving, Aja Woolley, Sophie McPherson, Sam Bjork, Madison Schultz,
Autumn Vine, Allison Lange

for Josh and Josie

C O N T E N T S

In science, the word 'error' does not carry the usual connotations of 'mistake' or 'blunder.' 'Error' in a scientific measurement means the inevitable uncertainty that attends all measurements. As such, errors are not mistakes; you cannot avoid them by being very careful.

—John R. Taylor, *An Introduction to Error Analysis*

MATERIALS

Resistance

Now you are a wind
across the hairline
balance of my scale.
I tare again; still

you draft past me.
The third decimal place
shudders under your weight.
I would never tolerate

uncertainty, the plus
or minus like apology,
a fraying at the edge
of measurement.

I used to want your signal
off my radar. Now
I'm savoring disturbance:
your Richter in my teeth,

the surge of current
in my elbow's circuitry.
The way I ran in socks
across the carpet, thrilled

with fright but powerless
to keep from reaching up
once more to catch
the doorknob's brass kiss.

Parasympathetic

Work to remember times when it didn't drop
or spill, when you caught hold of the glass before
it would have struck and shattered apart.
Think of the windows of Chartres, or concerts

Midori played with only a single E string.
Unbroken *chaînés*, and all the *changements* for which
your metatarsal stayed intact, and
thousands of grounders Bill Buckner fielded,

touching his toe to first without second thought.
Unharried mornings Anthony didn't have
to hear one prayer of yours, short-order
saint of disorganization, lost keys,

and pieces of things. All of the days we kissed
hello and goodbye, days when you didn't die.
That twisting last-minute back somersault,
body a blade and the landing dumbstruck.

Heart beats. How many perfect contractions, each
perfecter still than Mary Lou's dismount from
the pommel horse: the ten, ten
thousand times over again. And then.

Point Judith

Mornings of rocks thrown at the sea.
The rib-ache reaches toward the ridge
of shoulder blade, burrows under
collar bone, curls down the arm
hinge-honing, leaving only the heft
of denser stone, damp of salt-gnawed palms.

Men arrive with the tide, tilting
under poles and buckets, staking rod-holders
in the sand at prudent distances.
They linger, guarded over rig-tying,
stitch hooks through thick clam bellies,
crouched before the prayer of first cast.

Grains that glisten in the lifelines
and cling, even plunged clean.

Cicatrix

Found you that night with love
beating out of your shin, the wound
a dark potential of bone: open, oval
as an eye and yours wide at it.

Blood in beats, the throngs
of cells fleeing their halls and piling, rusted
at the edge of exit: the heme, dark rider,
steeled in oxidation.

Some part of me recalling
tibia before the rest reached out
and pinched the skin together
in a makeshift stitch—and this

the first time we had touched
in a week or more. Guests
arrived with bandages
like butterflies, while you

refused the hospital
(afraid since childhood) and held
its gaze, almost impressed.
Then looked at me, said, *Stay.*

We waited in the grass
between two torches
as long as harm
would keep us in its way.

Theorems for János

[1.1]

In the theory of parallels we are now not even farther than
Euclid. This is a shameful part of mathematics.
   —Carl Friedrich Gauss, 1813

Let me put it to you in a language
you can understand: the intersection
of you and me has become the null set.

You once said: Every person is a prime,
divisible only by themselves and one
other. To find that one is finding love.

Only—that division leaves you with yourself
intact. Exactly what I mean, you said.
But love's not that. It always gives back

a different number than you thought you had:
it's halved, or squared, or changed some way you can't
devise a function for. Love won't be

an identity. But you don't believe that.
You spend nights tracing lines from negative
to positive infinity, to prove

they never touch.

[1.2]

For God's sake, please give it up. Fear it no less than the sensual
passion, because it, too, may deprive you of all your leisure, your
health, your rest, and the whole happiness of your life.

    —Farkas Bolyai, letter to son János, on attempting a
    proof of the parallel postulate, 1820

So this is it, I tell them: this is all
he does.
                 Think of a line
and then a point not on that line, alone
in space, like a star.

             Now draw a parallel
(to the first line) through that star.

How many of those can you draw?

Just one—
              or more?

I tell you this not to belittle you
but so you know I know what it is that you do.

[1.2.1]

Even avuncular Euclid, looking over
your shoulder in dreams, would shy
from such a question. Smug,
axiomatic, he charts straight lines

from one point to another, finds their finite
segments, and proceeds to inscribe
circles of every center and radius. He sands
right angles down to size with earnest carpentry.

But even he—who fashioned these four postulates
to base and balance
his plane world—resents this fifth
and awkward leg—the wobble
at his table that allows a full
place-setting, but will not desist in its
unsteadying.

[1.3]

Out of nothing I have created a strange new universe.

> —János Bolyai, letter to father on formulating non-
> Euclidean geometry, 1823

And what if you found, or reasoned
into being, another realm
with its own rare, exacting rules?

Where people stood in arc
in the marketplace, the birds
ellipsing overhead— Would you still

dream each line processing
in its infinity of solitude,
unable to look up and see

the match that glides so near? Or this:
in the moment before you fall asleep,
at a single point, they kiss.

Spill

I have no dress except the one I wear every day. If you are going to be kind enough to give me one, please let it be practical and dark so that I can put it on afterwards to go to the laboratory.

—Marie Curie, on a proposed gift of a wedding dress

After all these years I will not wear
the gloves and coat, forget safety

glasses, there's an elegance to pouring
unprotected. Even the summer

I lost every mosquito bite
to the nitric bath, I felt grateful

to be etched clean as glass.
Nothing known can threaten here,

not poisons making themselves beautiful:
crystal violet, amaretto-scented cyanide,

ethidium bromide incarnadine.
This was sanctuary, graduated

by the keen meniscus, calculations
checked and checked again.

I liked that it left you out:
no trace under the hood, the waste

of charcoaled beakers and sulfuric acid
as I came and went each night

forgetting about my clothes, the soles
of my shoes, my fingertips to bread

and bread to lips until that day
I turned the Geiger on: the SOUND.

Lazarus

Trust I am sincere
as Martha and Mary ever were

Lord, if you had been here—
though we do not know,

says Socrates, that what we go to
is a lesser good. At that

our science loses frequency.
We can examine the body

once again returned to ground
among the nails they buried here.

Ockham says an afterlife
is too much lace around the neck

and you told me
if you believe

we are more than bags
of chemicals, you should not be here.

That was Introduction to Biochemistry,
or the hospital.

Melodrama

Hydrangeas and I'm back with Jimmy Stewart
gentlemanly averting his eyes as Donna Reed takes cover
in them. They are sexy like she is—downy, supple, plentiful,
spectrum from pink to blue, depending
(sorry Donna) on the pH of the soil. *Oh, hot dog!*
But when he says *I want to shake the dust*
of this crummy little town off my feet—and see the world!
I shift like the shapely blossoms, and when later
he throws the needle off the record and grabs her,
petals bursting between his knuckles, and holds the phone
away, her half-deaf mother carping upstairs—
she parts into his grasp blue and a darker gloom,
violet with fear and love of the future hold.
Play it, then. Do it from memory again.

Leaving the Theater

And when Hud, alone in that broken
down ranch house, puts down his hat
and turns to us with the open
beer in his hand, everyone dead

or gone away by then, and slams
the door in our waiting faces, we turn
to one another, gathering
our things and leave him there

with his beer in the house behind
THE END. Brushing against
each other's heavy coats, the voice
that in each of us says *Go on, get*—

moves us in droves toward the door.
What stalls? What keeps us here?

Americant

1.
I'm tired of you people saying I don't know what I'm doing.
Now it's not just the ones who were never gonna learn,
it's some of the rest. I told you I know what it's like,

I know you need to sleep at night, you need
to buy things for your family. I'm keeping that open.
You understand it's a turmoil here, we're in a struggle.

You know, it's a tug of war. Like the carnival,
and we've gotta keep the candy corn stocked. I mean,
you go expecting cotton candy, right? What happens

if someone turns mean on one of your kids?
You run him out of town. We've got an understanding.
Beans to bacon, I'm telling you. We're not gonna rest

until everyone's on the right side, till we've sorted it out.
If someone you knew got tangled up—
well, what were they doing over there?

We want to give you the opportunity to stay
as close as you can. We want you to flourish, and flower,
and in kind for you and yours. It's a proud thing, being a part.

2.
What I'm missing is a voice for me.
To say what I think, what I disagree.

Also a box to stuff my disappointments and movie stubs.
I'm sure I flubbed it when I introduced myself—

Maybe the answer is stop eating butter.
That one thing and I'm sure I'd look leaner.

I'm seriously angry about the wars, etc.
But the TV never listens. Lying here

I can feel the smoothness of the bat
and me up there. Then I cream it, right

in the sweet spot, you know? And it drives
over their heads, the wall, the park, it just goes.

3.
Bombs and black bombs at sunset,
at dawn. Your leader counted on
your distraction, and he was right.
Yes, he lied—who doesn't?

I haven't tried breathing in years.
My kids, they do it anyway,
like nestlings.

They're fearless with the street,
fire and car alarm and keening.
The bones of their skulls met in these sounds.

I go out to buy bread
regardless. Someday,
when all this is over, I'll make you some.

It will be in a restaurant, of course.
You'll smile and want to see into my soul.
You've told all your friends about this place.
Just sinking your fingers into it feels like atonement.

Rain

This notebook that sleeps beside me like an insecurity blanket,
the kitchen window underneath the slant of roof
that funnels the rain inconsolable,
the coffee pot I pour from even in dreams:
these are the elements.

Today the umbrella seemed an aegis
worth carrying forward across town,
some nobility to bear,
and these drops a constant—
interminable luxury of decimals
I could factor into an equation—
and I mean a linear one.

Didn't Noah and his sons, their wives,
get used to it after a while?
A moment when the terror subsided
into something they couldn't imagine
themselves without. Is it endurance or incapacity
that makes us, poor examining animals, keep—

Husmus

Mouse, we're wintering together.
The rats chased us up from the subway,
who can blame us? They've got
their whole city mapped out down there.

If I loved you, would you be so frightened?
I could coax your brown tremble out
from beneath the furnace, let you sally free.
Think how we could sing late into the night!

You're so studious and serious.
At night just my steps on the peeling linoleum
have you running for your book.
I know the cold helps you concentrate.

But what if we, what if we tumbled down
into the lentils together tonight?

METHODS

Inked

Once I thought it was measurable, love.
Like mercury
the beads cling to the insides of the pen
that I used to write you.
If I had been meticulous as Curie,
I would have weighed
the paper before and after, divided by
the specific gravity
of ink, and found a volume—microliters—
small enough
to ask for back. And if—

    Hah, you say.
You forgot to account
for oil, water,
the scent of your hand.

Sagres

8.
the end of the world. At the edge
we stood apart, then closer, looking out.

7.
the words I learned as a child. *Little Lamb*
who made thee / Dost thou know who made thee

6.
Three of them pushed us off as we called *O-*
brigados into the mist. I felt like one
thrown back.

5.
Portugal, you said, and it curled
and shimmered in the dark.

4.
and how we argued all night around
Sevilla. Every turn and we were back
at that same fountain. How I tore the map

3.
His hands swept toward each other until
the right one swerved the moment before

2.
He spoke in Spanish underwater. *Port-*
ugues? You shook your head. *Español?*
and he shook his.

1.
Screaming woke me. Mine. Fishermen
in yellow boots running to the car. I looked
at you and rolled the window down, and you
spoke Spanish to the man whose head reached through.

Gramercy

Near the park that took a key
to get in, three times a week

that summer, I waited for my boyfriend
on the steps of the halfway house.

After work, late light
signing the brownstone:

it was a beautiful halfway
house. When he came out

I could have waited longer.
Sometimes I came early

to do that, perfecting breathing
on those steps, sick with something

I still can't name.
People went in and out,

and I was jealous
of them—fallen

and picked up, picked up
and falling—those who opened

and held each other
responsible. Simple.

*The night I met her she drank more
and I drank less than ever before,*

he told me he told them,
laughing. I matched him

one for one, held his hand
as he spat up that night's trouble,

and the garden nearby
that took a lock to get in, it wasn't

my garden, and, he said,
there isn't a key.

Get Home Safe

Don't go to the party. Or go, but don't
talk to anyone. Leave early. On the subway

lock eyes with Dr. Zizmor, his magic
acne cure, or your copy of *Anna Karenina*.

When a man comes toward you on the long
walk home, cross the street. Actually,

don't cross; it makes you vulnerable.
Look him in the eye. Or seem

deranged, your tongue hanging out
the side of your mouth. As you pass him,

stick your keys between your fingers
like Wolverine. Hold your breath

until you're right across the street
from your door. When something flickers

just behind you, wonder why
you chose these shoes. Clatter across

and get hit, tenderly, by a car.
Execute the backwards somersault

from yoga class, stand up to face
the man who stops, picks up your book

and gives it back. You're right outside
your door now. No, he says, let me call

an ambulance. Get in, though nothing
is wrong. Someone is holding your hand.

Woman at the Well

She's holding up a piece of shiny plastic,
rasping, *Where's the bottle for this top?*

IMPATIENCE $10 A POT, barks the sign
outside the T. Red earmuffs over coffee,

Reverend Larry Love waits out April
jazzless. He has been the only clergy

in your life for quite some time. Missing:
his powder blue band uniform, the tall

white marching hat with pom-pom and elastic
chinstrap, the roller skates that wheel in spring!

Everyone who drinks here will be thirsty again.

Valentines for John Berryman

1.
Greedy I grabbed all the berries I could carry
and held them home till the stain wrote my palms.

2.
To get hold of the best, you have to pinch
between the thorns and pull back.
Get deeper in, stabbed and stuck,
it's punctured work the bloody berries.

3.
Like the ones that thrive under the overpass,
shade-taught. Your ear cocked between each carload, gone.
Don't be too difficult at first. Humble, mumble,
what's the point?

4.
You fisted it into a line, stuffed it in your grasp,
crunched, chomped like a bear. The roots come out
in your underwear.

5.
Sledding, the pricker bush dove at my lip
and gave me my first virgin loss
against the new snow wood.
Here you go, reenacting again. He falls for the—
how many times is it now? When will Veronica come?
Cherish is a word I use not lightly.

6.
The best berries will prick you up
before you get to them.
Upward stick 'em!
Then there's the odd one, waiting in the sun,
that no one noticed. Pick.
You don't it won't live longer.

7.
Rather fall than be eaten, rather dive than be saved.
Dove, dove, the wood awaits.

Housesitting

You had to turn down Swampy Dog Road.
An old New England farmhouse, set back

from the road. Too many books,
Moroccan rugs, furniture inherited

or acquired. A poster of Amelia Earhart:
Women Fly. That summer we took care of it

while your professor and her partner were away.
We fed the cats, watered the garden, slept

in their bed upstairs. I was allergic, but greedy
for the space and something else, which seemed,

despite their absence, so well tended. We picked
their basil and made pesto—the first time

I'd tasted it. You got up early
to surprise me with scones. Each day

I slipped my hand in blind and chose
three angels from the indigo bowl

on the bathroom sink, propped them
on the mirror lip: Kindness, Clarity,

Presence. Upstairs we thrilled each other
the way I imagined they did, until

the morning I wavered—ever a stray.
The stairs must have had a woven runner

I scampered down. I must have brushed
the rocking chair, the yellow Formica

table, the yawning back door.

Love and the Crumb Girl

The law that entropy always increases, holds, I think, the supreme position among the laws of Nature. If someone points out to you that your pet theory of the universe is in disagreement with Maxwell's equations—then so much the worse for Maxwell's equations. If it is found to be contradicted by observation—well, these experimentalists do bungle things sometimes. But if your theory is found to be against the second law of thermodynamics I can give you no hope; there is nothing for it but to collapse in deepest humiliation.

—Sir Arthur Stanley Eddington, *The Nature of the Physical World*

1.
She walked in and dropped her clothes
on top of other clothes already
on the floor and climbed
past them into bed. He had never seen

a mess so well-maintained.
And to be fair, it wasn't quite a mess,
it had achieved a certain echelon of chaos
such that it seemed enchanted, like the woods

in storybooks that vined into their own
while humans slept. But nothing like neglect:
the cups on saucers piled on books, their hinges
showing: such precariousness took skill,

or a talent at eluding basic principles:
there, that wine glass falling—
should have—still it hangs there
and she pulls him in.

2.
Afterwards, he thinks he wants to floss
his teeth—or, no: he doesn't
want to floss, he wants
to stay in bed like this with her

among the unlaundered ruins, half-
full water glasses. Let his gums keep
what's there beside them for a night.
At work they called her the Crumb Girl.

When he'd embraced her for the first time
earlier that night, he breathed in the scents
of her sweater—potatoes, carrots, onions,
wool—like the steam of good soup.

3.
She wasn't as cavalier about disorder
as she looked. And maybe, for the long-term,
that was good, but for the morning
he was disappointed. He didn't want her

up and out of bed, sorting it all
into piles, her back a question
mark, he needed her inside
the bed that floated, holy

and unsusceptible to the scatterings.
He saw now she was crying,
and he rose between two towers
of compact discs and tugged her back.

The way she navigated without even
glancing down he found disarming.
They lay all morning twined in silence,
no one knocking, calling.

4.
She lives like this, but mostly
by avoidance: she's hardly ever home.
Early she is gone past pigeons
with a morning of nodding ahead of them,

driveways returned to saltlicks,
to the job where she knows now
she's the Crumb Girl, he has told her,
wanting to make her laugh and bring

them closer. Not that it quite
bothers her, but now she feels
the image of herself tailgating
her: walk (echo), look up

(echo), stop at the cafe and order
the coffee she spills in small rhythms
across the floor, a tremor she
inherited from her grandfather,

and is usually proud of. Today
she feels slippery, the Person
Falling Apart: there's one
in every coffee shop.

5.
At Enviroveritas, she doesn't see him
until late. They're all too busy
getting ready for Mother's Day:
pushing bouquets aside

for Mother Earth, their appeal
to the family vote. *Mom's
running a fever!* shouts the placard,
where an oblate spheroid woman

convalesces with blue skin
and trees for hair. This is our best
attempt to stop doomsday-dealing?
The I-have-a-nightmare,

sea level rise, plagues, droughts, unless
we wise up and stop driving soon—
it's never worked. And yet, what other way?
She's writing the speech for the rally:

Electric light rail, solar,
wind! her sandwich
in one hand. Then *Saturday night?*
he's asking from the doorway.

She peers out from the sprawl
of papers, down at crumb-burrs.
Cringes. Glancing over at Mother
she smiles, chokes out *Sure.*

6.
Walking home, she thinks that at this moment
she would abide by any changes
required in the laws of physics
such that her scarf would stay on, such

that this was the first law of thermodynamics.
Yes, this would change a lot of things,
perhaps unpleasantly. But Saturday morning
she is feeling like a citizen. Her thighs will carry

her anywhere. Not home. To the tables
on the balcony of the Coop, perched over
the bestsellers, poetry and the thinning
hair of Harvard Square.

Set up with coffee, an afternoon
ahead of her. Thought can happen here.
Not in the enshambled room. But today
her neurons feel clipped, like trees or claws.

7.
There it is: the pamphlet called *The Warning
Signs of Squalor*. A woman on the street
just gave it to her. Did she know?
The warning signs of squalor: three of them

she recognized. It was a well-maintained (his
phrase) disorder. It was a propagation of not
having enough time (hers). It was a useful
fortress, dexterous warding away. *And what*

is powerless? they asked in the gymnasium
where she stood with her grandpa years ago,
trembling their Styrofoam cups. *Not until
you see someone else seeing it.*

So, she, before she met him. A propping up
of things that slouched from corners. Chaos
rearranged, but only a new look. She could
clean it up, she could. But that first law

insists on the conservation of mass
in the universe, or her apartment.
Or, by someone's theorem:
you can't win, you can only break even.

8.
Back home, she adds a new Neruda to the stack
she hardly sees, moves past it
as neutrinos do past us: neither partner noticing
the nearness of the swerve. He noticed.

He'd like to take notes, bring a caliper
to the error bars between her ankle and the piles,
how close she cuts it without disturbance.
Though it amuses him, it's Circle 9 for her.

Last stop, end of the line, everyone off.
Is disorder inherited or acquired?
Either way, the jeans have been draped
over chairs for generations.

Her grandmother's house approaches
geology or tree rings, layers
of newspaper sedimenting into shales.
Like watermarks the dates still bloom,

backwards. She imagines patient archaeologists
mapping out the site and sifting through it,
lifting with forceps the red beak
chipped off the rooster statue (*Portugal, 1953*),

gathering small white chunks of plastic
from the underlayer of the carpet,
tracing them by weight and size
to different snow globe manufactories.

9.
The pamphlet's warning signs: they weren't
what you'd expect, *e.g.* no: cockroaches,
ants, insects in general, stenches,
rodent visitations. This was squalor

defined as lifestyle, composed of *patterns
of avoidance* (their phrase, trademark).
You could be clean, beyond reproach
in one way yet still be living in squalor.

First of the patterns: going home
only to sleep. Next, spending days
in third places: bookstores,
cafes, even (after these have shut)

grocery stores, wandering, not buying.
Number three they called the Goldilocks:
making yourself too much at home
in other people's spaces.

10.
The pamphlet takes liberties she doesn't
approve of: they call disorder *entropy*,
which, she understands, is jargon (*squalor*
isn't spot-on either), but the lack of rigor

gets to her. Regardless, quoth the pamphlet:
Unless you take daily steps
to combat entropy, it will
overtake you. She wouldn't quarrel

with them—the second law
enforces that: *You can't even*
break even. She reads testimonies
about the moment something snapped:

Pauline A., violinist, missed
a concert when she couldn't find
her other black high heel.
The one that makes her think

this may not all be bunk
is the lengthy history by poet Arthur G.,
how the weeks without
his buried Larkin almost broke him.

11.
They chase each other from the lab-bench bar top
at the Miracle of Science back to his place
in Central Square. She walks in, and
Pythagoras, inside her head, could mark

and calculate all angles between points
where furniture touches down. Books
on shelves, coats on hooks, shoes in rows
by the door. He takes her elbow,

guides her to the middle of the floor, and draws
a circle around them in chalk, then drops
the contents of his pockets. She thinks
of Marie Curie, and wants to tell him

how when she and Pierre finally found
that enormous signal firing from coarse
pitchblende, she brought in dump trucks
of the stuff, her passion in the isolation

and measurement of this radium—
to have it pure enough to plumb
its most immediate dimensions,
as if discovery were second to accuracy.

12.
Even Odysseus and Penelope, after exhausting
the night, Athena stopping time, let dawn
sift silence between them, sleep part them
like a curtain. The sleep of single minds.

When she wakes, she discovers the book
next to his bed, parked a perfect parallel
to the nightstand edge. She flips
to the crisply nested bookmark:

A careless shoe-string, in whose tie
I see a wild civility:
Do more bewitch me than when art
Is too precise in every part.

RESULTS

Old Maids

Once we shared a mouse. He darted from the corners
of our eyes; he curled up in our lentils. We were too tender
with him, but it was winter. We felt for him
within our creatures, our morning aches, night hunger.

Then he found our dreams—yours baby mice
(*Oh, boys or girls?* your mother asked), mine a blind
and injured one I couldn't catch. Our ancestors
had wired us not to keep close quarters. But habits

of our own—inclined toward corners, to be left alone,
each of our brown caps bent over a book—
made war hard to declare. We started missing
patches of our underwear. Our friends

put up a scoreboard: ever nil for us
and more for mouse, they questioned theories
of a single shooter. I read over your shoulder
the family of mouse takes six weeks to quintuple.

That tripped some timorous switch in us. We pitched
our stakes—I can't say how—on human ground, until
we heard no feet beneath the stove, no doughty
prodigy sprung from the cooling toaster.

We met men, we moved away. Eventually
we married them. One winter afternoon
I tell your sons about it—the quiver
catching up with me. How once I loved a mouse.

Thoughts and Prayers

The white pine fell
the afternoon before
the newest war began.
Unlike the day before
there was no wind
though it was uncharitably cold.
The tree sacked
our power line and halved
a car we later learned
had also been a home
but whose owner, mercifully, was out.
We heard the robot zeugma
of our pins to all pressure points
in the galaxy blink out,
then the transformer pop.
I climbed into bed and read.
The house got cold. Of course
I coddled the battery
of my phone like an injured wren.
The neighbors volleyed news
and rumors of restoration.
Some of us went out to examine
the downed god, broad
as a fridge and moments before
a thumbprint's width from heaven.
The arborist pronounced mold
in the roots. Impossible to foresee
except for a subtle yellowing
of the needles. The owner thought
it may have been its time,
pine beetles, climate change.
We ordered pizza, told jokes
around the lantern. The toddler
tried every light switch in the house.
We went to bed early. I could hear

the neighbors inside their houses,
talking softly. No one hurt,
thank God. Old silver being polished
back to life. You fell asleep.
I listened.

False Spring, Year of the Tiger

Red couplets of the new year bright
around my neighbor's blue door

climb up over the hill into sun
still there, 3 p.m. Presidents' Day

another false leader proclaiming land
his own his birthright west

wind licked off the Sound
combs the bare cottonwoods flat

and shivers me cottonmind cotton-
heart Spring comes early in Seattle

wrote the *Times* reporter who visited
for a weekend and left us

someone's flag caught on the roseless
thornbush the pinwheel we got

at the beach last summer says wind
coming from all directions or none

buffeted stand in this patch of sun as long
uncertainty levels high and growing

even in dreams precision and accuracy
are not the same the sound I thought

was the rain was the wind in the pines
in the road dozens of spruce cones we've

forgotten how to windbreak each other
boughing to what in you is me all along

my walk No Cleaning Job Too Big
or Small when I think of a joke to tell you

and remember I can't   high in the cedar
a creaking fledgling or branch about to

Moscow, Vietnam, mindfulness
in the miniature library and the 100-year-old

sequoia the neighbor called just a baby
hand to bark we're still learning

Postcards of the Numbered Empire

Sanctuary is finding the right size.

Can you bow low enough
to pass through the door,
climb up to the covered balconies

below the three-tiered roof
of the synagogue in Zabłudów—
larch wood, not one nail—

that students in Algiers
glued back together by centimeters
nine years after the Nazis burned it?

Meanwhile he sits, c. 1890,
in feathered warbonnet,
unsmiling, thumb pinching middle finger,

surrounded by seven white women
smocked and contented
with their miniature sculptures of him.

I stretch apart
my thumb and pointer,
trying to see.

Today or yesterday
the President frenzied thousands,
call and response.

What was the name?
Animals.

Big Muskie dug
until he couldn't. The world's hugest
excavator scrapped in 1999

except for his 12-car-garage-sized
bucket
they saved for the tourists.

My Chemistry

I used to follow parcels of water
around the ocean
a student called them *parasols*
which makes perhaps
as much sense as a box of brine
someone like Santa
or the UPS man might deliver from North
Atlantic into depths
to flow southward toward a fate
in calm Indian
or eggbeater swirl of Southern
Ocean around Antarctica

I lifted molar fingerprints
to trail parcels
suffused with their last gulp of surface—
CFCs marking time
with their singular blend in the
atmosphere each year:
here's a fine '77, bottled off Greenland,
now worming through
Titanic (these compounds hermetic,
their signatures blurred
only when a container of fridges is made to
walk the plank),

or carbon-14, which keeps a
memory of bomb
for any parcel surfacing since 1950
revealing how much
has been swallowed and sunk by
its own cold
saltiness, how long it takes
to crawl through
four oceans and float back up—1,000

years, and what
will it wake to, a hothouse or snowball,
how far will

we have gone or be gone by then,
and what'll you
do without us, we wonder like a capsized
lover, the way
I heard the whales in the Bay of Fundy
had stress hormones
lower than scientists ever measured before
in late September
2001, quiet among millions of invisible floaters
carried by currents
and giant fecal pellets that must frighten them and the sea
snot or mucilage

plankton might exude in heat although
there are not
many experiments you can do down there,
better to be
an observationalist—detecting what is or trying to
write it down
and not get too seasick along the way: I once
played badminton on
a container ship's deck with the captain and kept
hitting the birdie
on top of the containers he would smile and say
it's okay I

have another one until finally after I had done it
again and again
he said *I think it is best if we stop playing now*
that is how
science feels most of the time that is my
chemistry: a parcel,
something to hold in your hands,

bang skull against,
tearing and unfurling its carapace, each
cubic-metric Davy
Jones' Locker full of swarms of molecules and ions
flowing in and

out of all six sides along the gradients of salt oxygen
carbon dioxide silica
from which diatoms make their
slim glass parasols
to enchant the microscope slide,
one box after
another, piled high and wide, an ocean
of these boxes
like the screen full of small ones
full of numbers
I gathered while at sea and took back
to my room

to turn into wild
pictures of deep
feeling (for a time)
that they were mine

Animalia Pacifica

1. Estuary Logic

Just before sunset the otter
sleek and river-combed
crept up on the rock
to dismantle a crab

c l a w s   f l a i l i n g

I heard the crunch
and crack as he eyed me   ate
then his darkness
became the water's

2. Hum

It's that squeaky scritching
sends me searching for him—
more insect than bird—
persistent, whittling the air.

Stand still long enough and spot
the silhouette of bobbin and spindle,
or bar dart needling a higher branch,
short king announcing spring.

Is it fear or thrill that starts him
speed-diving down, playing wind
through fluted wings
then soaring up again—a joke,

Juliet! All my emeralds and rubies
I lay at your throat.

3. Reunion

The crows gather there:
the parking lot
behind the university gym
each afternoon, they say,
because it used to be
a burial ground,
then a dump
overlooking Union Bay.

After the Cut
was dredged, the lake
fell 9 feet, birthing
small peaty islands,
one named Broken.

I skirt the lot as they converge,
shrill and cross, searching for the passage
one canoe wide
before the scars and maps
and burns, the floating bridge,
the city thick with distance.

Three Salmon Redactions

redact: from Latin redigere: *'to bring back' or 'to reduce'*

1. Mortal

In U.S. Pacific Northwest coho salmon (*Oncorhynchus kisutch*), stormwater exposure annually causes unexplained acute mortality **when** adult salmon migrate to urban creeks to reproduce. By investigating this phenomenon, **we** identified a highly toxic quinone transformation product of *N*-(1,3-dimethylbutyl)-*N*'-phenyl-p-phenylenediamine (6PPD), a globally ubiquitous **tire** rubber antioxidant. Retrospective analysis of representative roadway runoff and stormwater-affected creeks **of** the U.S. West Coast indicated **widespread** occurrence of 6PPD-quinone (<0.3 to 19 micrograms per liter) at toxic concentrations (median **lethal** concentration of 0.8 ± 0.16 micrograms per liter). These **results** reveal unanticipated risks of 6PPD antioxidants to an aquatic species **and** imply toxicological relevance for **dissipated** tire rubber residues.

2. Home scent

Elevated concentrations of CO_2 in seawater can disrupt numerous sensory systems in marine fish. This is of particular concern for Pacific salmon because they rely on olfaction during all aspects of their life including during their homing migrations from the ocean back to their natal streams. We investigated the effects of elevated seawater CO_2 on coho salmon (*Oncorhynchus kisutch*) olfactory-mediated behavior, neural signaling, and gene expression within the peripheral and central olfactory system. Ocean-phase coho salmon were exposed to three levels of CO_2, ranging from those currently found in ambient marine water to projected future levels. Juvenile coho salmon exposed to elevated CO_2 levels for 2 weeks no longer avoided a skin extract odor that elicited avoidance responses in coho salmon maintained in ambient CO_2 seawater. Exposure to these elevated CO_2 levels did not alter odor signaling in the olfactory epithelium, but did induce significant changes in signaling within the olfactory bulb. RNA-Seq analysis of olfactory tissues revealed extensive disruption in expression of genes involved in neuronal signaling within the olfactory bulb of salmon exposed to elevated CO_2, with lesser impacts on gene expression in the olfactory rosettes. The disruption in olfactory bulb gene pathways included genes associated with GABA signaling and maintenance of ion balance within bulbar neurons. Our results indicate that ocean-phase coho salmon exposed to elevated CO_2 can experience significant behavioral impairments likely driven by alteration in higher-order neural signal processing within the olfactory bulb. Our study demonstrates that anadromous fish such as salmon may share a sensitivity to rising CO_2 levels with obligate marine species suggesting a more wide-scale ecological impact of ocean acidification.

3. Elwha

The removal of two large dams on the Elwha River was completed in 2014 with **a** goal of restoring anadromous salmonid populations. Using observations from ongoing **field** studies, we compiled a timeline **of** migratory **fish** passage upstream of each dam. We also used spatially **continuous** snorkeling surveys in consecutive years before (2007, 2008) and after (2018, 2019) dam removal during summer baseflow to assess changes in fish distribution and density over 65 km of the mainstem Elwha River. Before dam removal, anadromous fishes were limited to the 7.9 km section of river downstream of Elwha Dam, potamodromous species could not migrate throughout the river system, and resident trout were the most **abundant** species. After dam removal, there was **rapid** passage into areas upstream of Elwha Dam, with 8 anadromous species (**Chinook, Coho, Sockeye, Pink, Chum, Winter** Steelhead, **Summer** Steelhead, Pacific Lamprey, and Bull Trout) observed within 2.5 years. All of these **runs** except Chum Salmon were also observed in upper Elwha **upstream** of Glines Canyon Dam within 5 years. The spatial extent of fish passage by adult Chinook Salmon and Summer Steelhead increased by 50 km and 60 km, respectively, after dam removal. Adult Chinook Salmon densities in some previously inaccessible reaches in the middle section of the river exceeded the highest densities observed in the lower section of the river prior to dam removal. The large number (>100) of adult Summer Steelhead in the upper river after dam removal was notable because it was among the rarest anadromous species in the Elwha River prior to dam removal. The spatial extent of trout and Bull Trout remained unchanged after dam removal, but their total abundance increased and their highest densities shifted from the lower 25 km of the river to the upper 40 km. Our results show that **reconnecting** the Elwha River through dam removal provided fish access to portions of the watershed that had been blocked for **nearly** a century.

Arctic Circle

Small Christmas bulbs that cast a prickly light,
the wooden booths, a traveling violinist,

ice, the plain of ice, a team of dogs,
their cries, the wind, a shout, then light

applause, a guitarist joins the violin,
someone in the corner blushes, hands

touch hands, a blade scrapes golden
life-bearing fat, the dog-fur strands

of the hood freeze at their edges,
chocolate clings to a fork that rests

on the demi-plate, and the village girls
reach up to compare the curve of breasts

of strangers with those of their mothers,
glacier swallows another boulder blue.

Letter to Martian

There is a substance we cannot live without
that for a moment gets lighter as it gets colder.
Because of this, creatures can live beneath it in winter.

People tell more truth the farther they get
from home. This is why we long to travel.

The substance surrounds us in many forms
and keeps us strange to one another. We fear it
though we float on it.

When other creatures speak to us
we cannot understand, but we love them

and kill them and praise their bravery in war.
Many wait in the street for summer
and when it comes it is too hot.

We cannot shake the feeling someone
is responsible, though we have no evidence.

In case you could hear music
we send it soaring away from us.
We search for water.

Gwaxčeł

Once a weir
trapped salmon where
this summer my daughter
picks salmonberries

The creek softens
ice-forged drumlins
culvert-sprung
to where the people
of the big lake wintered

silent place

I try to hold it in crow memory
what was
what has settled
what has never been

DISCUSSION

Question

It's an old family story, how at five
I asked, and they turned
the question back on me. *Well,*
where do you think you came from?

I replied, they say, as cool as milk,
From earth and from God.
I find it hard to believe now. But then, I was
still close enough to both

to know, I guess.

Before the Funeral

The orange juice is grey. My father keeps
calling me Mary. (His youngest sister's name.)
Every five minutes the newsprint swirls
beneath his eyes. We look away, out
the kitchen windows to the aftertaste
of snow along the sidewalk. Gravel, rock
salt, and the rust-edged grass between the cracks
like stubble. My brother slurps the purpled milk
left on his cereal spoon. Nobody tells him
to stop. My mother moves between us,
fastening. I reach down and touch the place
where the nail polish has hardened to my leg,
to stop the run. I have been caressing it—
for how long now? My fingertip is numb.

Meeting Robert Frost Behind
Montebello Elementary School

We waited in the woods for him
the day our teacher said he'd visit.
Thin woods behind the school the builders
had passed over, why did he want
to meet us there, where a stream gathered courage
in springtime and we dared each other
to smell the tongues of skunk cabbage?
Now dry leaves underfoot, the birches
leafless. It was cold and he
was late. Fierce rouge smudges
defending her cheekbones, our teacher said
he wasn't coming. He was dead.
She took out a tape recorder,
one of the old rectangular ones,
and laid it on a boulder. Then
the crotchety woodsman spoke to us,
as if with the stump of something—
a cigar, a bit of wood—at the corner
of his mouth. Half attentive to something else,
not trusting us. Some of us cried,
out in the darkening woods so deep
in the season, as he kept going.

To E.D.

Marsh House, Amherst, Mass.

What they called your eccentricity
made sense to me, the way
the parable of talents
didn't. He wasn't hiding his gifts,

that servant cast into outer darkness
to wail and gnash his teeth,
he was preserving them.
I never questioned your solitude.

But what if a century bowed
and excused itself from the space
between us? Out my window
I look through the trees through yours.

Far neighbor, could any word of mine unveil
the cautious wingspan of your smile?

Waking

Along this seam someone is basting, someone is ripping stitches.
Branches. At their edges, light particular and past blue gathers.
Afternoon already older. Branches thicken into tree
somewhere below and only then: fugue clouds,
how fast they plunged across the sky and left
when someone says that word, the word for you,
says it,
louder,
again.

Salt Pond, 1989

Restless in the slip,
the bleached Boston whaler can't decide, keeps
my brother busy counting minnows deep.
It's the slap of sun that starts us
down the muddy bank for release, sunk
between hungry swans. We lope salt-licked
over barnacles and twisted butts, the slough
of older kids' evenings. That summer I slid
out the window to walk the dark slant
to the water—or so I wrote—but never
had the recklessness to help myself.

But all is to be dared

One day I realize I have the power
I have always longed for: when I sense
I am not needed in a scene—
my mother and her sisters,
my lover and an old friend—I disappear,
extinguish, unnoticed.

                    Only, matter
has its way—I can't entirely vanish.
I have to be *something*—and remain.
Whatever is easiest. A letter opener. A spiral
notebook. But every time spills ink, breaks plate,
sticks door. I'm trying to be still here.

                    But I'm bog
their knees sink into, snifter his nose snubs,
napkin she lipsticks.
Sappho understood—*greener than grass*—
excuse yourself hungrily, then flood.
They laugh and wipe it up.

Calving

1.
The years become one wait
two weeks at a time.

The drugs balloon me into
what I hope to be but won't.

Things peed out by pregnant women,
things like the hair of an asp

Puck must find for Oberon.
Every try they show us a snapshot

of the best blast, with the corona
of cells that will feed it.

It is itself
and its own nourishment

until it isn't.

2.
Inside their blubber orcas carry
old pollutants—PCBs, flame

retardants, pesticides: the ones
with bonds that break

for no one, stick around—
though mothers have the lowest levels

because they pass them to their calves
in milk. Sunset, I walk around the park

among children,
mothers, the memory patients.

An older woman, off-kilter
but friendly, says *STOP*

and points up at the dark red clots.

3.
My favorite nurse at the clinic
returns from maternity leave.

We're still here. There's something
funny about that, right?

Does your body still feel
like a science experiment?

asks one of my only friends
who knows. There are tricks,

pangs and tucks and pings
in my abdomen, a furry feeling,

the body fucking with me.

4.
How does it feel? To hope and guard
against it like Mary and Martha

waiting up for Jesus.
Like middle-aged Elizabeth

surprised by John the Baptist
dancing within her.

You must be ready for the miracle
and the chance absorbed again,

as the petals form another year.

5.
I no longer dream I have abandoned a child,
no longer dream about children at all.

Or daydream
about names or roller skates

or just a molten heap
in my lap. Potato sack.

Or flour, the five-pound bag
we had to carry around in junior high.

I was so confident then.
I named my flour sack Aurora Chloe.

Once, in a rush, I left her in my locker.

6.
Once all I wanted
was time to make,

didn't need a little
nudging into me

didn't need more family—
then this rumble

late in life:
fear of death?

or just the urge
to bless someone

now that I've had time—
your brand-new name

(less fragile
than your body now)

inventing its own light.

Receptivity Assay

Today the painprint still wet.
You have been lucky. The doctor said,
your eyes musseled shut,
When I was in medical school

they told us women can't feel
anything inside the uterus.
A pause. *They were wrong.*
That plain blue bolt, it split

the soul and body. Not
like the time you got hit
by the bicycle, the time
you got hit by the car.

Or when you fell on the marble
steps, or—gingerly—took a scalpel
to your wrist. Body cries out
before asking how brave

you'd like to look. Soul comes back
to scorn: How lucky you have been.
She rests the box of tissues
on your abdomen.

Sacré-Cœur

To stay on the horse,
look to the left or right
of the blinding.

On the second day
at the Opéra stop
a woman stood and gestured to her seat.
Years of wishing and falling
short, I shook my head and took it.

The first of three days
harboring uncertainty
I hauled myself up

the steep hindquarters
of Montmartre—wide steps, uneven—
dead end of pastis-colored houses
and to the church
where I lit a candle not for you

but for my grandmother
who late in life
dreamed of a little pied-à-terre,

walking to Mass each morning.
You, then, a cluster of promise
some portion of which
had bloomed red and damp
on landing. I thought: *Of course.*

Errors in the nuclear code.
Down the culvert.
Waited a beat too long.

Didn't want enough.
Still: how I fled the stifling
famous bookstore, cooled myself
with boules of fraise des bois
and pêche de vigne, could not escape

the scent of smoke.
Three days abiding
wilderness, mind canted:

living partwise—lest I upset
wave or particle,
relic or spark.
Then in the office
behind a stone garland

a doctor named Valentine
turned a dial, and the small white room
filled with thundering hooves.

Dream Feed

It shouldn't hurt
it burns
she throttles the right nipple
pimple that has always been there
poem takes shape in the dark
one finger   my dimmed phone
O muse

I never thought I'd have
stopped daring
hope
Remora
in a fleece angel suit
Empress Josephine
in Elizabeth's ermine robes
4 a.m. you feed
white noise spatters behind

blue glow you smile at   old friend
all the emails I haven't returned and
the ones I have only to hear nothing
all of us
thin skeins

Spelunker you come up in the dark
cheeks shining
arms triumphant
I pull you to me
living person of my body
I never thought this would be
for me

I Walk Through the Neighborhood Without You for the First Time Since Your Birth

Camellia blossoms brown at the edges already

I will be 60 when you are grown

my mother a week ago silent after she told me this

doing her own math

I cried so she didn't have to

you will have let me go by the time you are that age

cherry blossoms   wedding rice

already 8 o'clock it's getting dim—

birds calling to nest

porch lights on

blossoms sweet on the wetcool

the pungent dryer vent says all is well

passing Benjamin Bunny and Zuzu Bear

and a girl whining to her mother

I'm only having a bad day because of you

Benjamin fat and yellow can't tell I'm crying

but black sleek Zuzu eyes me skeptically

camellias already rubble, first flowers of spring

some brown low

some still pink corsages

granny bushes

I once disparaged them—

How Hard We Try

Sliding the glassware into the autoclave,
I thought about it, how we do our best
to scald ourselves clean for the next
generation, but there's always the gunk
at the bottom that gets baked on.

Could we return to when we first
understood each other?
That day in the hospital you said
I hope you know what to do
and I latched on.

Porosity

Season of salmonberry then   currant
raspberry
       thimbleberry   cherry   almost
blackberry

of ants in the kitchen

of napping while she naps
writing undercover

the blanket naked

                            (its crimson sleeve
                            whirling in the wash)

On this morning's walk with Josie
a dog named Sedona
a thousand whys

Summer's unboundaries pour us &

I wonder if my neighbor is angry
or worse.

The ants come marching in
the kitchen windows

Out back where Peggy's ashes
settled at Easter

her pale pink roses
trumpeting.

Will this be how I teach Josie
about death—or when I wipe the ants up
with a sponge?

(We had an unusually wet spring.)

The neighbor's irritation marches over
the soft pink tones of his wife
and daughter.

(She lived in this house
almost all her life.)

Each day the sun shines, the trees ripple,
I walk all the way to the park,
I am holy

(weeks
I prayed *restore my bellows*
feared
my life retracted)

so what escapes now is let in
unquestioned,
like a breath

weaving
alveoli   i l o v e a l (l)
interstitial   i startle in it
heal

rasp
thimb
sal
straw
black

sirens bagpiping up

(imagine Josie furrowing
I hope someone is okay)

to be spared for another rinse
another tumble
tongue bunched with fruit
from her palm

Hunter's Moon

The moon showed her whole face to you—
more than you can say for most of us
this year, eclipsed by masks—

and you haven't stopped calling for her
since. On Halloween, a harvest blue,
she gazed down warm and solemn

on your wolf suit. Next day we stretched
our clocks like bowstrings back
to claim the waning light for larks,

so for a while you slept too soon
to see her rise. Still, on afternoon-
becoming-evening walks

you crooned her name while we
reconstructed dusty orbits, blamed clouds
or rain when it seemed easier—

how much had we lost hold of, the years
we ate in yellow kitchen light?
You settled for the curve of my

half-satellites, yet once you found her
in the morning, pulling Sound
full up on sand, and reached out

crying *Own*. And on a late walk
as I pondered full or empty
you pronounced her simply *D*.

You offered her your pacifier,
later you proffered water,
which she declined, already

having some. Now moon
is apple crescent, fluff of lint
from winter sheets, your face

tranquil then electrified.
Last night she seemed a smudge,
a thumbprint behind clouds, until

we went outside and—full again—
she silvered out to center stage.
Josie, no one can own the moon,

but tonight I offer her
your brow, asking her to bathe
the whole of you.

Myth

You're telling me Orpheus played a mean lyre.
So mean he convinced the gods to let him
bring his love (her name was Eurydice)
back from the dead. And the only catch
is he can't look back at her
till they're out of there. Easy, right?
So they're hoofing it up the switchbacks of hell,
they're doing all right. They stop to rest
and he sticks to his guns. But just as they round the bend
of the underworld for good, he thinks
maybe he'll sneak a peek. They'd never notice.
So he turns—he turns and glances back.

                                        And what?

So—it turns out he only looked back at his collar.
You grab my whole leg in your arm. He never
actually saw her! Your mouth
comes over mine. So they're fine.
They are living together even now.

Valentine

Leaning over the kitchen sink,
T-shirt damp and soapy at my belly,
I once said, these dishes
are like us. The white plates with blue trim
that almost match, though one's a little smaller,

the speckled bowls I bought
at Shiga's on the Ave, feeling rich,
the spoons with slender handles
my mother gave me when I left home.

Sturdy, familiar, motley, modest.
Filled and emptied daily, splattered
and restored in a piping baptism.
Good as new, eternal—then dried
and stowed together in the dark.

We have a dishwasher now.
One hairline fracture healed
miraculously in its geyser.
The plates have chipped

but never broken.
That day you said nothing
but enveloped me,
your tall trunk warm and forever
against my back.

No Smoke, No Heat, No Rain

Porcupine quills of the silver pine
needle the rainbow scroll

of dusk, late July: an eyelid dusted
bluepinkwhite over the vague Cascades—

distant fires... Car alarm and harmonica
of Lake City Way, Rocky barking

and the neighbors watering their plum trees.
How many more summers like this,

what we used to call summer?
To breathe the air and sit in sleeves

hearing seeing smelling
night come on. My husband always

goes in earlier—*getting eaten up*—
as I linger, spared for now, nursing

that fir in the distance, each bough shifting
like a cat's tail at sea—even more at the top

where one stout trunk becomes two.
Birds knit the last light with nest talk

then silence before the traffic picks up
and someone whistles invisibly.

NOTES

"Theorems for János": At age 18, the 19th-century Hungarian mathematician János Bolyai discovered non-Euclidean geometry by showing that Euclid's fifth postulate, the parallel postulate, could not be proven from the first four. It was a problem mathematicians, including Bolyai's father, Farkas, had been trying to solve since Euclid's time.

The epigraph in "Spill" is from *Madame Curie: A Biography* (1937) by Eve Curie, translated by Vincent Sheean.

"Lazarus" echoes part of Martha's statement to Jesus in John 11:21.

"Sagres" includes lines from William Blake's "The Lamb."

The ending of "Woman at the Well" references Jesus's words in John 4:13.

"Love and the Crumb Girl" references Ginsberg's theorem and ends with lines from Robert Herrick's "Delight in Disorder."

"Old Maids" is for Rebecca Starks.

"Postcards of the Numbered Empire" was inspired by postcards from the Musée d'Art et d'Histoire du Judaïsme and the Art Institute of Chicago shared by Emily Warn. The poem quotes chants from Donald Trump's rally in Nashville in May 2018. He urged the crowd to call MS-13 members animals, a word he has also repeatedly used to refer to immigrants.

Sources for "Three Salmon Redactions":

"Mortal": From Tian, Zhenyu et al. "A Ubiquitous Tire Rubber–Derived Chemical Induces Acute Mortality in Coho Salmon." *Science* vol. 371,6525 (2021): 185–189. doi:10.1126/science.abd6951. Reprinted with permission from AAAS.

"Home scent": From Williams, Chase R et al. "Elevated CO_2 Impairs Olfactory-Mediated Neural and Behavioral Responses and Gene Expression in Ocean-Phase Coho Salmon (Oncorhynchus kisutch)." *Global Change Biology* vol. 25,3 (2019): 963–977. doi:10.1111/gcb.14532.

"Elwha": From Duda, Jeffrey J et al. "Reconnecting the Elwha River: Spatial Patterns of Fish Response to Dam Removal." *Frontiers in Ecology and Evolution* vol. 9,765488 (2021): 1–17, doi:10.3389/fevo.2021.765488.

"Gwaxčeł" is a Lushootseed word meaning "go for a stroll/walk there," according to a translation by Warren KingGeorge, historian for the Muckleshoot Indian Tribe. KingGeorge proposed this name for a part of the Thornton Creek watershed, which inspired the poem.

The title and a line from "But all is to be dared" come from Sappho's Fragment 31, translated by Anne Carson.

ACKNOWLEDGMENTS

I thank the editors and publishers of the following journals, anthologies and projects in which these poems were first published, some in earlier versions:

32 Poems: "Gramercy" and "Get Home Safe"

4Culture: Poetry in Public: Places of Landing: "Gwaxčeł"

America: "Question"

Bennington Review: "Thoughts and Prayers"

Boxcar Poetry Review: "Cicatrix"

Cider Press Review: "Calving"

DIAGRAM: "Lazarus" and "Spill"

EcoTheo Review: "Old Maids"

House Mountain Review: "How Hard We Try" and "Waking"

I Sing the Salmon Home: Poems from Washington State: "Three Salmon Redactions"

The Merton Seasonal: "Parasympathetic"

Moist Poetry Journal: "Porosity"

Moss: "My Chemistry"

Mud Season Review: "Point Judith," "Americant," "Valentines for John Berryman," "Before the Funeral," "Woman at the Well"

Pacifica Literary Review: "Housesitting"

Pidgeonholes: "Sagres"

Plant-Human Quarterly: "I Walk Through the Neighborhood Without You for the First Time Since Your Birth" and "No Smoke, No Heat, No Rain"

Poetry Northwest: "Husmus"

Pontoon Poetry: "Resistance"

Psaltery and Lyre: "False Spring, Year of the Tiger"

Radar Poetry: "Sacré-Cœur"

Revel: "Meeting Robert Frost Behind Montebello Elementary School" and "Salt Pond, 1989"

Seneca Review: "Theorems for János"

The Shore: "Letter to Martian"

Tahoma Literary Review: "Love and the Crumb Girl"

The Threepenny Review: "Leaving the Theater"

The Yale Review: "Melodrama"

* * *

Thank you to Dr. Ross Tangedal, Sam Bjork, Eva Nielsen, and everyone at Cornerstone Press for your kindness and care with my book. Thank you to Elaina Ellis for your transformational editing.

Thank you to Eunice Kim for allowing me to include your sublime print on the cover.

Thank you to Rebecca Starks for so many years of deep friendship and literary collaboration.

Thank you to Laura Da', Keetje Kuipers, Ed Skoog, and Rosanna Warren for your beautiful words and encouragement.

Thank you to my teachers, Rosanna Warren, Robert Pinsky, Glyn Maxwell, and April Bernard. Thank you to all the incredible poets in my cohort at Boston University who imprinted me with your wisdom.

Thank you to the Fulbright Program, Hugo House and Laura Lampton Scott, the Elizabeth George Foundation, and Artist Trust for supporting my work.

Thank you to Cyndy Hayward and Jeff McMahon at Willapa Bay AiR, Danielle Epstein, Dina Schapiro and Marnie Briggs at Marble House Project, and Carrie Hardison at Sitka Center for Art and Ecology, for the gift of time, sustenance and focus, nurturing and supporting my artistic development in such beautiful places. Thanks also to all the inspiring writers and artists I was so fortunate to be in community with during these residencies.

Thank you to all my friends and family near and far for your support and love, particularly Tim Aubry, Jeri Helen Belisle, Eliza Calhoun, Spencer Hamblen, Rafe Jones, Letty Limbach, Anne Nester, Andy, Winslow, Malcolm, and Lionel Solomon, and Jeremy Sussman.

Thank you to my parents, Peggy and Charlie Lockwood, for your endless love and support, and for always encouraging my creativity. Thanks especially to my mom for introducing me to poetry so early and convincing the leader of a poetry workshop at Yosemite National Park to let me in when I was 5.

Thanks to my brother, Christopher Lockwood, for your support and belief in me. Thanks to Melissa Tiedge and Jacob and Olivia Lockwood.

Thanks to my Washington state family: Matt Distler, Becca Fishaut, Julia and Calla Distler, Elinor Distler, Phillip Ross, Tom Distler, Phyllis MacCameron and Shane, Anna, Rían, and Finn McHugh.

In loving memory of my grandparents, Margaret and Gerald Shea and Matilda and John Lockwood.

Thank you to Josh and Josie, my world.

DEIRDRE LOCKWOOD is a Seattle-based poet, fiction writer, and journalist. Her poetry has appeared in *The Threepenny Review, Yale Review, Poetry Northwest,* and elsewhere, and her journalism has been published in *Scientific American, Nature, Science, Chemical & Engineering News,* and the *Chicago Tribune.* She has received support from the Fulbright program, Hugo House, the Elizabeth George Foundation, Artist Trust, Marble House Project, Willapa Bay AiR, and Sitka Center for Art and Ecology. She holds an M.A. in creative writing from Boston University and a Ph.D. in oceanography from the University of Washington.

www.ingramcontent.com/pod-product-compliance
Lightning Source LLC
Chambersburg PA
CBHW031436120626
46545CB00006B/2425